Guy de maxence Afanda

Mathematics reconsiderations

THE REMAINING CHANGE

The residual variation of the quantity x or the variable x is:

$$\Delta_n x = \frac{x^{n+1} - x_0^{n+1}}{x^n}$$

Residual differential is then: $d_n x = (n+1)dx$

Each time, n is the frequency; it is what -measured in "time" - maintains the shake.

There is steady state at: $\Delta_n x = 0$. The following table illustrates it:

n	x
0	x_0
1	$x_0 , -x_0$
2	$x_0 , x^2 + x_0 x + x_0^2 = 0$
3	$-x_0^2 , x_0^2$
...	...

If f is a function or application, the stationary motion obeys the relation:

$$\Delta_n f(x) = 0$$, with:

$$\Delta_n f(x) = \frac{f^{n+1}(x + \Delta x) - f^n(x)}{f^n(x + \Delta x)}$$

$$\equiv \frac{f^{n+1}(x) - f^n(x_0)}{f^n(x)}$$

Stationarity then requires: $x_0 \neq 0$, and respectively: $f(x_0) \neq 0$.

These studies consider the resounding actions these that remain after the extinction or disappearance of the cause. In this case, with a function or application f there is resounding action if: $df = \varphi f dg + \Phi dg$, with: $\varphi \neq 0$, and: $\Phi \neq 0$, where g is the directory function or application, and φ and Φ are rhythms.

We will note that in this case the derivative of the null function 0_f is   not   necessarily   zero. Thus,
$$d^2 f = (\varphi' + \varphi^2) f dg^2 + (\varphi\Phi + \Phi') dg^2$$; such that:
$$\begin{cases} \varphi' + \varphi^2 : \text{the oscillator} \\ \varphi\Phi + \Phi' : \text{the resounder} \end{cases}$$; f is a passing function or application.

7

An example is f defined by:

$$f(x) = x^{n+1} + (-x)^{n+1} = [x^n - (-x)^n]x \; ; n \text{ is a}$$

frequency.

The study of this function calls some reconsiderations we will do now. This is the definition of the logarithm function. Basically we have: $F(x) = \int \dfrac{dx}{x}$

Then turn:

. $F(ab) = F(a) + F(b)$

. $F\left(\dfrac{a}{b}\right) = F(a) - F(b)$

. $F(x^p) = pF(x); p \in R$

. $F(2^x) = x$; that writing is done with respect to the binaring formation of integers numbers; Indeed each integer is written in the form: $N = \sum\limits_{i=0}^{m} \alpha_i 2^i$; $\alpha_i \in \{0,1\}$.

Generally, f $(x) = \sum\limits_{i=0}^{n} (-1)^{E(i)} \dfrac{x^{n+1}}{n+1!} f^{(i)}(x)$; with:

$f^{(i)}(x) = \dfrac{d^i f(x)}{dx^i}$; n is the nomial degree of

8

f. By application: $\int \frac{dx}{x} = 1 + \frac{1}{2} + \frac{1}{3} + ... + \frac{1}{n+1}$; it appears that it is well suited to write that: $\int \frac{dx}{x} =$ Logx=Log2^y=y, in this new case. Also, a $\leq Logx < b$, such that: Logx=a+k2^{-a}, with: k $\in \{n \in N/0 \leq n < 2^a, a \in N, b \in N\}$.

By studying the function f defined by: f(n)=$(-1)^n$= $2^{nLog(-1)}$, it appears that Logx is defined algebraically for x>0 (algebraic logarithms). So, it should be recognized that Log(-1) is given geometry (geometric logarithm), ie, the size of the area such as: $\int_{4}^{2} \frac{dx}{xLogx}$.

THE HETEROGENEOUS DIFFERENTIAL

They are of the form: $$\sum \varphi_i df_i(x_i) = df(x_1,...,x_n)$$

For example, let: adf(x) + bdg(y) = 0 ; it is equivalent to: af'(x)dx + bg'(y)dy = 0. For the resolution, proceed as follows: ask, A = af'(x), B = bg'(y) , such that:

Bdy + Adx = 0 (1); So:

dAdx+ dBdy = 0 (2).

By exploiting the equation (1), we get: BdA - AdB = 0; is: A = BC^{te} .

ALGORITHM TO SOLVE THE PROBLEM OF n BODIES

Remember that : $\dfrac{a}{b} = \dfrac{c}{d} = \dfrac{a+c}{b+d}$. Going further, we have: $\dfrac{a}{b} = \dfrac{c}{d} = \alpha\dfrac{ac}{bd}$; we deduce:

$$\alpha = \dfrac{d}{c} = \dfrac{b}{a} = \dfrac{b+d}{a+c} \text{ ; from where: } \dfrac{a}{b} = \dfrac{c}{d} = \left(\dfrac{b+d}{a+c}\right)\dfrac{ac}{bd}$$

.

In this example, get:

$$\dfrac{a}{b} = \dfrac{c}{d} = \dfrac{e}{f} = \left(\dfrac{df+bf+bd}{ac+ae+ce}\right)\dfrac{ace}{bdf}$$. More broadly,

we have: $\dfrac{x_i}{y_i} = ... = \dfrac{x_n}{y_n} = \dfrac{\Pi_{n-1}^{y_i}\Pi x_i}{\Pi_{n-1}^{x_i}\Pi y_i}$; note that for

example: $\Pi_2^{(a,b,c)} = ab + ac + bc$ (the multiplication of a b c two by two).

Now consider two bodies of respective masses m and m' in remote interaction. They act in ways equal one to the other according to the necessary equality of action and reaction. The mutually exerted forces have same

13

intensity, such that: F = F '=

$$\varphi \frac{m}{d} = \varphi \frac{m'}{d'} = (\frac{\varphi d + \varphi' d'}{m + m'}) \frac{mm'}{d^2}$$

Based on experience, we need:
$$\frac{\varphi d + \varphi' d}{m + m'} = C^{te}$$

With three bodies we have:

$$F_1 = \frac{m_1 v_1^2 \sqrt{d_1^2 + d_1'^2 - d_1 d_2 \cos \alpha_1}}{d_{1d_2} \sin \alpha_1} = \varphi_1 \frac{m_1}{d_1 d_1'}$$

; therefore:

$$F_1 = F_2 = F_3 = C^{te} \frac{m_1 m_2 m_3}{d_1^2 d_2^2 d_3^2}$$

$$F_1 = F_2 = ... = F_n = C^{te} \frac{\prod m_i}{C_2^n \prod_{i=1}^{} d_i^2}$$

For n body, like:

THE ROUND CARTESIAN MARK SYSTEM

The flat Cartesian coordinate system is the one usually used either the mark whose matrix form is: $\begin{bmatrix} 1 & 0 \\ 0 & 1 \end{bmatrix}$, as the matrix or formal canon of Cartesian orthonormal coordinate system (O, $\vec{i}, \vec{j}$) ; thus, a position point (x; y) in this frame is in this position precisely. Rather, the matrix: $\begin{bmatrix} 0 & 1 \\ 1 & 0 \end{bmatrix}$ is the matrix barrel or the matrix form of the reverse mark system (O,$\vec{j}, \vec{i}$), such that a position of point (x, y) in the reference position has rather the position (y; x). Thus it should be noted that the transition from a mark sytem to a reverse mark

system is a rotation of angle α such that: $\alpha = \dfrac{\pi|x-y|}{2\sqrt{x^2+y^2}}$;
$(O, \vec{j}, \vec{i})$ is the round Cartesian mark system in this case.

In the space stage, the matrix barrel mark is: $\begin{bmatrix} 1 & 0 & 0 \\ 0 & 1 & 0 \\ 0 & 0 & 1 \end{bmatrix}$; and reverse mark system or round mark system is matrix base: $\begin{bmatrix} 0 & 1 & 1 \\ 1 & 0 & 1 \\ 1 & 1 & 0 \end{bmatrix}$, such as the position (x, y, z) of a point in the flat space or direct reference $(O, \vec{i}, \vec{j}, \vec{k})$ rather is:

$(y + z, x + z, x + y)$ in the reverse mark system $(O, \vec{k}, \vec{j}, \vec{i})$. There is thus a rotation of angle θ such that:

$$sin^2\theta = \frac{r^2 - (R-r)^2}{r^2} = \frac{2Rr - r^2}{r^2} \; ;$$

$$r = \sqrt{x^2 + y^2 + z^2} \; ;$$

$$R = \sqrt{(x+z)^2 + (x+y)^2 + (y+z)^2} \; .$$

An algorithm to find the average radius is: being an ellipse whose large radius is A and the smaller radius a, trigonometry   of   the   ellipse   is:

$$\begin{cases} u\sin\alpha = (a-u)\cos\alpha \\ v\cos\alpha = (A-v)\sin\alpha \end{cases}$$

$$u = a - R\sin\alpha\,; \quad v = A - R\cos\alpha\,;$$

And if A is taken as a basis, R is the mean radius; the calculation gives:

$$R^2 = \frac{A^2 + a^2 tg^2\alpha}{1 + tg^2\alpha} = a^2\sin^2\alpha + A^2\cos^2\alpha$$

THE DUALISTIC MULTIPLICATION

This state of affairs requires the following clarifications:

1°) 1≠ +1; that is, x ≠ +x ,in the sense that if x indicates a quantity or magnitude, +x indicates a transaction or

$$+2= \int_{1}^{3} dx, \quad -4 = \int_{6}^{2} dx,$$

course; e.g.:   +2= 1   6   are paths; +x indicates the increment or the arrival, -x shows the decrement or the removal ; therefore the operation:

   1-2 = 0-1, gives different result of the operation:

+1-2 = (+ 1) - (+ 2) = +0-1 ; So if 0 indicates nothing or absence, +0 = -0 , indicates inertia or conservation; then 0-1 = rev(1-0) = rev(1); +1 reverse is -1, but the reverse of 1 is 1,"1 crossed" ,a hole or a void or absence. In contrast, +0-1 = -1.

2) We must distinguish the discontinuous, digital, or arithmetical multiplication, and continuous or geometrical multiplication; the arithmetical multiplication is this:

a ++ b = ab ; the geometrical multiplication is as follows:

$$a \times b = \begin{bmatrix} a & a\sqrt{1 - \sin \alpha} \\ b\sqrt{1 - \sin \alpha} & b \end{bmatrix};$$

$\alpha = \hat{ab}$; the argument of : $a \times b$,is:

$$\arg (a \times b) = \det (a \times b) = ab\sin \alpha .$$

New considerations to keep then appear necessarily in the form of discriminatory ratings and specific ways of doing contrary operations.

For ratings, we establish in turn:

i. $x ++ x = x^2$; and   therefore:

$x ++ ... ++ x = x^n$

ii. $x \times x = x^{+2}$; and therefore:

$x \times ... \times x = x^{+n}$.

for contrary operations, we are called to distinguish distributive division and fractal or sequential division one hand, the induced square root and the geometrical square root on the other. So in turn:

1) Distribution

We formulate it as: $a -- b$ (a reduced b times); when it is equal, we have known writing: a = bq + r ; when inequality sets, we have: $a = \sum\limits_{i=1}^{b} q_i$.

In general, you have already noted that precisely, a = $q_n b + r_n 10^{-n}$,where n is the number of completed subdivisions; for example, $11 -- 4 = 2,75$; Here, there are two subdivisions; ($r_0 = 3$, $q_0 = 2$), ($r_1 = 2$, $q_1 = 2.7$) and ($r_2 = 0$, $q_2 = 2.75$); also consider the following example: $11 -- 7 = 1,5714285...$; we consequently analyze the following table:

n	r_n	q_n
o	4	1
1	5	1.5
2	1	1.57

21

3	3	1,571
4	2	1.5714
5	6	1.57142
6	4	1.571428
7	5	1.5714285

In truth, this division is equal if there exists r_n such that $r_n = 0$. Otherwise, we must admit that:

$D = \Sigma q_i = \Sigma(q + \delta_i) = qd$; therefore: $\delta_i = q_i - q$, and: $\Sigma\delta_i = 0$; So $: D -- \ d = q$, is egalitarian, while: $D -- \ d \in \{q | D = \Sigma q_i = \Sigma(q_0 + \delta_i) = q_0 d\}$, is unequal. As a result, 2-1 and 2+1 for example, are unequal halves of 4.

2) Sequential division

If for distributive division quotient is defined as "... for each" (the part of everyone), sequential division result is frequency: "...times" . For examples: $3 -- \ 2 = 1,5$ "for each" or 2 and 1 "for each respectively"; 2

$$\frac{3}{2} = 1,5$$

"time"; by application : $c = a \times b$, leads to:

$$a = \frac{c}{b} = \det(c) -- b\sin \hat{b}a$$

; in   general:

$$z = \frac{x}{y} = \det(x) -- y\sin \hat{y}z$$

3) The induced square root

In fact, the divisibility is the ability to perform a finite number of sub-divisions after division.

The set of real numbers is the set of positions or extents on a straight half-line containing an origin O; they are divisible for the continuity of the half-line, mainly by two, according to the duomerical formation of integer numbers.

Any amount not belonging to a straight half-line taken as referential set, is a bynumber or geometric frame, namely the result of a deformation of the right line; for example, $\sqrt{2}$ is the result of: $(1+1)^{\widetilde{\frac{\pi}{2}}}$, that is $(1 + 1)$ modulated by the deformation $\frac{\pi}{2}$; indeed, observe the following figure:

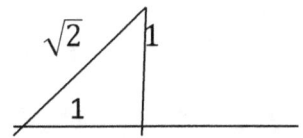

So in general, compared to a ray taken as reference:

$$(a+b) \sim \alpha = \sqrt{a^2 + b^2 - 2ab\cos\alpha}$$

$$(a-b) \sim \alpha = \sqrt{a^2 + b^2 + 2ab\cos\alpha}$$

$$(a++b) \sim \alpha = \begin{vmatrix} a\cos\alpha & a\sin\alpha \\ b\sin\alpha & b\cos\alpha \end{vmatrix} \tag{1}$$

$$(a--b) \sim \alpha = \begin{vmatrix} a\cos\alpha & a\sin\alpha \\ \sin\alpha -- b & \cos\alpha -- b \end{vmatrix}$$

$$(a \times b) \sim \alpha = \begin{bmatrix} a & a\sqrt{1-\sin\theta\cos\alpha} \\ b\sqrt{1-\sin\theta\cos\alpha} & b \end{bmatrix}$$

Thus,
$$\sqrt{2} = (1-1) \sim \frac{\pi}{2} = (1+1) \sim \frac{\pi}{2}$$, and in general: a $\in N$,

$$\sqrt{a} = \sqrt{i^2 + j^2} = (i+j) \sim \frac{\pi}{2}$$

Further:
$$\sqrt[n]{a} = \sqrt{i^{2(n-1)} + j^{2(n-1)}} = (i^{n-1} + j^{n-1}) \sim \frac{\pi}{2}.$$

4) The geometrical square root

It is noted : $\sqrt[+]{x}$, where x is an area. Thus:
$\sqrt[+]{x} = \sqrt[+]{a \times b} = \hat{a},b$, the angled juxtaposition of a and b.

Along,with:

$$a \times b \times c = \begin{bmatrix} a & a\sqrt{1 - \sin \alpha_{ab}} & a\sqrt{1 - \sin \alpha_{ac}} \\ b\sqrt{1 - \sin \alpha_{ab}} & b & b\sqrt{1 - \sin \alpha_{bc}} \\ c\sqrt{1 - \sin \alpha_{ac}} & c\sqrt{1 - \sin \alpha_{bc}} & c \end{bmatrix}$$

And with: $\alpha(\alpha_1,\alpha_2,...,\alpha_n)$, a multi-angle, we get:

$$\sin (\alpha_1,\alpha_2,...,\alpha_n) = 1 + (n-1)\prod\sqrt{1 - \sin \alpha_i} + \sum \sin \alpha_i - n$$

$$\sqrt[+n]{x_1 \times x_2 \times \cdots \times x_n} = x_1,\hat{...},x_n .$$

*

Relation (1) shows that: $(a ++ b)\sim\dfrac{\pi}{2} =- ab$; so that:
$(x ++ x)\sim\dfrac{\pi}{2} =- x^2$.

In other words, the equation: $x^2 + a^2 = 0$, shows that:
$x^2 = (a ++ a)\sim\dfrac{\pi}{2}$, then: x = a⁻ , the side of a deduced

25

square. Moreover, this equation leads to:

$$\begin{vmatrix} x & -a \\ a & x \end{vmatrix} = \begin{vmatrix} a & -a \\ a & -a \end{vmatrix}$$

Therefore, it emerges both: x = a and x = -a ; that is, x= $\pm a$, enantiomeric number or dinumber . As well,

x^2 = (+a)++(-a)= -a^2 .

THE PERMEABILITY

Ω is an object of thickness ε relative to a grounded area basis; there is pore multiplication if: $\displaystyle\sum_{\epsilon_i \leq \epsilon} o_i = qA$; there is pore superposition if:

$$\sum o_i = \int_{\epsilon}^{\epsilon_i \leq \epsilon} A d\epsilon$$

The object Ω is permeable if it includes a tunnel or tunnels several large smoothness. The tunnel is the total

$$\sum o_i = \int_\epsilon^0 A d\epsilon = o_\lambda$$

overlay of pores; So: , is a tunnel. The tunnel length is the shortest distance, through the tunnel. The tunnel is right if: $l_o = \epsilon$; thus, if the roughness of the tunnel is r such that:

dr= $-\dfrac{dp}{p}$, where p is the smoothness of the tunnel, so:

$p = \epsilon\, 2^{-r}$, minimum permeability of Ω is: $\pi_\lambda = \lambda\, 2^{-r}$, where λ is running in the tunnel. The overall

$$\pi_\lambda = \lambda \sum 2^{-r_i} > \int_0^\epsilon (2 - 2^{-r}) ds$$

permeability is: .